BE A WASTE WARRIOR!

SCHOOL WARRIOR

GOING GREEN

by Claudia Martin
Consultant: David Hawksett, BSc

BEARPORT
PUBLISHING

Minneapolis, Minnesota

Editor: Sarah Eason
Proofreader: Jennifer Sanderson
Designer: Paul Myerscough
Illustrator: Jessica Moon
Picture Researcher: Rachel Blount

Library of Congress Cataloging-in-Publication Data is available at www.loc.gov or upon request from the publisher.

ISBN: 978-1-64747-698-4 (hardcover)
ISBN: 978-1-64747-705-9 (paperback)
ISBN: 978-1-64747-712-7 (ebook)

For more information, write to Bearport Publishing, 5357 Penn Avenue South, Minneapolis, MN 55419. Printed in the United States of America.

CONTENTS

THE BATTLE TO SAVE EARTH!

People create a lot of waste—especially at schools. It's not only all the paper and pen waste from classrooms, but also the food and packaging waste from the cafeteria. But you can help battle school waste by making a few simple changes. Get ready to tackle school waste, Waste Warrior!

The Three Problems with Waste

Heaps of Garbage A lot of garbage ends up in **landfills**. When waste breaks down in landfills it can harm our environment. For instance, when **plastic** breaks down, harmful materials can leak into the soil and air. Waste in landfills is squashed together, which keeps out oxygen. So the waste rots slowly, and as it does so, it gives off harmful amounts of methane gas. Although most modern landfills are designed to stop harmful materials and gas from leaking out, the waste is still there—and being stored for a future generation to deal with. That is why waste warriors avoid creating waste!

We have created landfills to deal with our huge amounts of waste.

Wasted Resources When we get rid of something, we might need to buy something new to replace it. For example, if we get rid of used paper at school, we need more paper to write on. This means more trees must be cut down, wasting our planet's precious **natural resources**.

Polluted Planet We create **pollution** when we burn **fossil fuels** to power the factories that make things—including school supplies. When **fuels** such as **coal** and **oil** are burned, they release **carbon dioxide** and other gases that trap the sun's heat around Earth. This is causing rising temperatures and changing weather conditions around the world.

Air pollution from factories is a huge cause of climate change.

Trees are being cut down without being replaced in a process called **deforestation**.

The Six Rs

So you think it's time to battle school waste? You can use these six weapons in the battle to save Earth. Don't worry if you can't master them all. Be kind to yourself. Keeping just one piece of waste out of a landfill is a step in the right direction!

A key skill for a waste warrior is to know which materials can be recycled.

Refuse If you have a choice, try to avoid products wrapped in unnecessary packaging.

Reduce Try to get new products, such as pens and backpacks, only when you truly need them.

Reuse Before throwing away products or packaging, consider how they might be used again.

Repair If it's possible to fix something that is broken, do that. You may need to ask an adult for help.

Recycle Send unwanted products and packaging made from glass, paper, metal, and plastics to a **recycling** plant.

Rot Put your food and garden waste in a **compost** bin so it will rot away.

Ripped old T-shirts can be upcycled into book covers, pencil cases, and bags.

Used paper and cardboard could get a
second life as decorations
for a plant pot.

SAVE PAPER

More than half of the waste made by schools is paper. There are old books, stacks of notes, completed tests, and more! And what's worse, much of this paper ends up in landfills.

Over 10 percent of the waste in landfills is paper or cardboard. And every year, more trees are cut down to make paper—enough to cover the entire state of Maryland in a forest. Sometimes, new trees are planted to replace the ones being cut down, but not all the paper used in schools comes from forests that have been grown in this way. By being thoughtful with paper, a waste warrior can help save trees and reduce the paper in landfills.

An average elementary school student creates around 44 pounds (20 kg) of paper waste every year.

Every year, about 15 billion trees are cut down around the world.

What a Waste!

Around one billion trees' worth of paper is thrown away every year in the United States.

A great way to stop more trees from being cut down is to recycle paper whenever you can. Used paper can be recycled into new paper, greeting cards, or even toilet paper rolls! However, paper can only be recycled up to seven times, so trying to use less paper is also a good idea. Remember to write on both sides of paper and reuse waste paper for brainstorming.

Warriors Can Try:

Ask your teacher if you could try these paper-saving plans.

- Decorate old cardboard boxes to use as recycling bins for scrap paper.

- Every week, pick a classmate who will be in charge of collecting waste paper.

- Spread the word! Tell your friends and classmates about the importance of paper recycling.

Recycling is one of the six Rs. It's an easy way to help the planet.

PAPER

RECYCLE

Paper is recycled into paper or cardboard of lower quality. Writing paper often becomes packaging, and packaging becomes toilet paper.

PICK YOUR PENS

Plastic pens are a school staple. Most pens are **disposable**, which means they are designed to be thrown away when their ink is used up. But this creates waste! So what's a writing waste warrior to do?

People used to write with pens that could be refilled with new ink. Disposable pens became popular starting in the late twentieth century. Disposable pens are cheaper, but they are often made in a way that makes them hard to recycle. To recycle pens, which are made of a mixture of plastics and metals, these materials must first be separated from one another. Unfortunately, this means pens can't be thrown in curbside recycling bins, so there are now billions of pens in landfills.

Disposable pens usually have caps, bodies, and ink tubes made of plastic, while the tips are made of metal.

Billions of ballpoint pens have been sold since they were first invented in 1931.

What a Waste!

In the United States, 1.6 billion disposable pens are thrown away every year.

The simplest way to reduce pen waste is to use a pencil when possible. Pencil shavings can go in a compost bin, and pencils wear away over time, which means there is very little waste in the end. When you need to use a pen, look for pens made from recycled plastic, such as used plastic bottles. These pens reduce the amount of new plastic being produced!

Warriors Can Try:

Take your pick of these low-waste writing tools.

- Wooden pencils
- Pencils made from recycled paper
- Pens made from recycled plastic
- Refillable pens

This low-waste pen is made from recycled plastic and cork.

REUSE RECYCLE REDUCE

Pencils are made of mostly wood, so they will rot faster than plastic.

Have you ever noticed how many of your school supplies are made of plastic? Think about the plastic-covered folders, notebooks, pencil sharpeners, and pencil cases. Plastic can be difficult to recycle, so it poses a waste problem.

There are many different types of plastic—most of them are made from petroleum, an oil. This oil is a fossil fuel, formed over millions of years from dead plants and animals. We are using oil at a faster rate than it can form, which means it is a natural resource that we risk running out of. Some plastics can be recycled by being melted and reshaped. Others can only be ground up and used for packaging. While recycling plastic is good, it isn't the perfect solution to our school supply waste woes.

Since the first mass-produced plastic was invented in 1907, the material has taken over our world—and pencil cases!

What a Waste!

Every year, about 10 percent of the world's oil supply is used to make plastic.

PVC plastic from items such as clear folders can only be ground up and used for packaging.

For a wise waste warrior, finding an alternative to buying plastic school supplies does not need to be expensive or difficult. The first place to look is your own home. Check for supplies left over from last year! Or ask your family for items they no longer need. Reusing supplies is always better than buying new. If you do need something new, consider whether a nonplastic alternative will work.

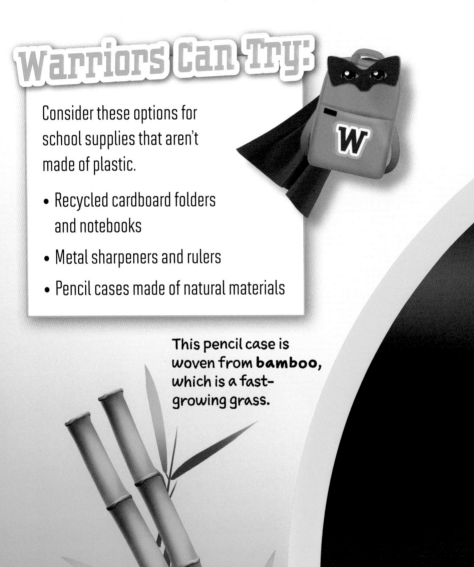

Warriors Can Try:

Consider these options for school supplies that aren't made of plastic.

- Recycled cardboard folders and notebooks
- Metal sharpeners and rulers
- Pencil cases made of natural materials

This pencil case is woven from **bamboo**, which is a fast-growing grass.

A wooden pencil box will last for many years. If it breaks, it can be left to rot.

LIMIT LUNCH WASTE

Packed lunches can create waste because of the wrappers and bags we use to keep food fresh. A bag for your bologna, a wrapper for your wrap, or a carton for your cookies can be bad for Earth.

A key problem causing lunchtime waste is the **single-use** plastic used to pack sandwiches and other foods. Single-use items are intended to be used just once and then thrown away. Even if sandwich bags only spend a short time in your lunchbox, they can take 1,000 years to **biodegrade**. As a result, sandwich bags pile up in landfills. Yet a waste warrior doesn't need to go hungry! Small steps can add up to a big solution.

Sandwich bags are often made of a plastic called polyethylene.

What a Waste!

Every year, the average American family uses 1,000 sandwich bags.

Since they are lightweight, sandwich bags can easily blow or wash into rivers and lakes, where they are a danger to fish and other animals.

The first step in reducing sandwich bag waste is to find an alternative. Discuss this issue with whoever buys your groceries. Ask if you could swap single-use sandwich bags for reusable options. Although these sometimes cost more to buy, they can be used year after year, possibly saving money in the end. Also consider packing foods that don't need wrapping, such as fruits and veggies.

Warriors Can Try:

To reduce the waste of single-use sandwich bags, think about these alternatives.

- Washable sandwich bags or containers
- Biodegradable sandwich bags made from plant materials
- Reused plastic packaging from bread or other grocery items
- Reused plastic tubs from spreads and other items

Reusing packaging year after year means you are not creating waste.

Paper coated in beeswax
keeps sandwiches fresh.
Plus, it can often be reused!

START A SCHOOL GARDEN

From potatoes to corn to bananas, much of the food we eat has been transported across the country or even the world. Trucks, planes, and boats that move our food are usually powered by fuels made from oil, which releases carbon dioxide when it is burned.

The extra carbon dioxide in our air is trapping the sun's heat around Earth. As a result, over the last century, the temperature of Earth's air and oceans has increased by about 2 degrees Fahrenheit (1 degree Celsius). This has been enough to melt sea ice, cause more storms and floods, and change habitats for plants and animals. For example, polar bears hunt the seals they eat on sea ice, which is now melting. But a school garden is a great place for a team of waste warriors to start making a real difference.

The fruits and vegetables you eat may have arrived on big container ships like this one.

what a Waste!

In the United States, the average meal has traveled about 1,500 miles (2,424 km) to get from where it is grown to our plates.

The rice in your school lunch may have come all the way from Asia.

Dole

The first step toward starting a school garden is to bring your concerns to your teachers. If your school doesn't have room for a garden, maybe you can grow seeds on a windowsill or plant fruits and veggies in containers on the sidewalk or playground. You and your **eco-friendly** classmates can volunteer to plant, water, and weed.

Warriors Can Try:

Follow these steps for planning your garden.

- Choose a sunny spot where it is also easy to water.

- Plan what you will plant during each season.

- Ask local farms, garden centers, or stores for donations.

- Set up a plan to divide the work. Maybe you and your classmates can take turns!

Carrots are ready for harvesting three to four months after planting the seeds.

You can start small
with your school garden.
It doesn't need to be big to
be a step toward change.

Make a Worm Compost Bin

Turn your school's food waste into compost with a worm bin! Compost stores the goodness from the foods it is made from, which means it can be used to help new plants grow. This is a perfect project for a green-thumbed waste warrior.

You will need:

- A 5– to 10–gallon (19 to 38 L) plastic bin with a lid
- An adult helper
- A drill
- Newspaper
- A spray bottle of water
- Red wiggler worms (available online)
- Food scraps

1 Ask your teacher or parent if you can make a worm compost bin.

2 Ask an adult to drill small holes in the sides of your plastic bin, so the worms have air.

3 Rip newspaper into strips and fill the bin halfway with newspaper. Spray the strips with water. This will keep your worms damp.

4 Add red wiggler worms to your bin. (Ordinary earthworms don't do well living indoors.)

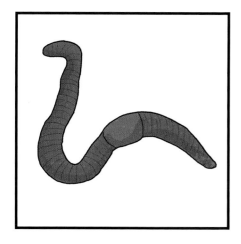

5 Place food scraps, such as fruit and vegetable peels, among the newspaper. Be sure to skip meat, oily foods, and dairy products.

6 Check your worms and food scraps regularly to make sure they are not completely dried out or so wet that liquid is washing around. Add food waste every few days.

7 After three or four months, you will start to see castings. Castings are worm poop, and they are rich in nutrients from the food the worms have eaten. Add the castings to your plants or school garden to help new food grow from old food!

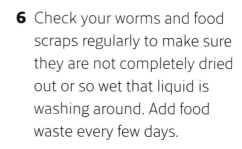

Glossary

bamboo a quickly growing, tall grass with a woody stem

biodegrade to be broken down by living things, such as bacteria and fungi

carbon dioxide an invisible gas in the air that is released when coal and oil are burned

coal a solid fuel that is found in the ground, made from the remains of animals and plants that lived long ago

compost rotted plants and food that can be used to feed soil

deforestation permanently clearing the trees from a wide area

disposable intended to be thrown away after being used

eco-friendly good for the environment

fossil fuels fuels made from the remains of animals and plants that lived long ago

fuels materials that can be burned to make heat or power machines

landfills pits where waste is dumped then covered with soil

natural resources materials found in nature, such as trees, water, metals, and coal

oil also called petroleum; a liquid fuel that is found in the ground and is made from the remains of dead animals and plants

plastic a human-made material, usually made from oil, that can be shaped when soft, then sets to be hard or flexible

pollution any harmful material that is put into the ground, air, or water

PVC a plastic that is often used to make clothing, packaging, and folders

recycling collecting, sorting, and treating waste to turn it into materials that can be used again

single-use an item that is meant to be used only once and then thrown away

upcycled recycled into a new product with a different use

Read More

Andrus, Aubre. *The Plastic Problem: 60 Small Ways to Reduce Waste and Save the Earth.* Oakland, CA: Lonely Planet, 2020.

Brown, Renata. *Gardening Lab for Kids.* Beverly, MA: Quarry Books, 2014.

Poynter, Dougie. *Plastic Sucks! You Can Make a Difference.* New York: Macmillan Children's Books, 2019.

Wassner Flynn, Sarah. *This Book Stinks! Gross Garbage, Rotten Rubbish, and the Science of Trash.* Washington D.C.: National Geographic Children's Books, 2017.

Learn More Online

1. Go to **www.factsurfer.com**
2. Enter "**School Warrior**" into the search box.
3. Click on the cover of this book to see a list of websites.

Index